SECRET

MARITIME UNIT FIELD MANUAL —

STRATEGIC SERVICES
(Provisional)

Prepared under direction of
The Director of Strategic Services

SECRET

SECRET

MARITIME UNIT FIELD MANUAL
– STRATEGIC SERVICES
(Provisional)

Strategic Services Field Manual No. 7

SECRET

SECRET

Office of Strategic Services

Washington, D. C.

18 July 1944

This Provisional Basic Field Manual for Maritime Unit is made available for the information and guidance of selected personnel and will be used as the basic doctrine for Strategic Services training for the operations of these groups.

The contents of this manual should be carefully controlled and should not be allowed to come into unauthorized hands. The manual will not be taken to advance bases.

AR 380—5, pertaining to the handling of secret documents, will be complied with in the handling of this manual.

William J. Donovan

Director

TABLE OF CONTENTS

SECTION I—INTRODUCTION

1. SCOPE AND PURPOSE OF MANUAL . . 1
2. DEFINITIONS 1

SECTION II—OPERATIONS AND METHODS

3. MISSIONS 2
4. CLANDESTINE FERRYING 2
5. MARITIME SABOTAGE 3
6. MILITARY TACTICAL ASSISTANCE . . 3
7. SPECIAL TRAINING BY MU 3
8. EQUIPMENT 4

SECTION III—ORGANIZATION AND PLANNING

9. BRANCH AND FIELD BASE ORGANIZATION 4
10. PLANNING AND ORGANIZATION FOR OPERATIONS 5

SECTION IV—PERSONNEL

11. REQUIREMENTS 6
12. RECRUITING 6

SECTION V—TRAINING

13. BASIC TRAINING 6
14. SPECIALIZED TRAINING . . . 7

APPENDIX "A"—EXAMPLES OF TYPICAL MU OPERATIONS

1. INTRODUCTION 8
2. MARITIME SABOTAGE (1) . . . 8
3. MARITIME SABOTAGE (2) . 9
4. CLANDESTINE FERRYING (1) . 9
5. CLANDESTINE FERRYING (2) . . . 10
6. CLANDESTINE FERRYING (3) . . 10

APPENDIX "B" -

1. MU OVERVIEW 13

SECRET

MARITIME UNIT FIELD MANUAL STRATEGIC SERVICES

(Provisional)

SECTION I—INTRODUCTION

1. *SCOPE AND PURPOSE OF THE MANUAL*

This manual sets forth the authorized functions, operational plans, methods, and organization of Maritime Units (MU) as a part of OSS operations. Its purpose is to guide Strategic Services personnel responsible for planning, training, and operations in the proper employment of Maritime Units.

2. *DEFINITIONS*

a. OVER-ALL PROGRAM FOR STRATEGIC SERVICES ACTIVITIES—a collection of objectives, in order of priority (importance) within a theater or area.

b. OBJECTIVE—a main or controlling goal for accomplishment within a theater or area by Strategic Services as set forth in an Over-All Program.

c. SPECIAL PROGRAM FOR STRATEGIC SERVICES ACTIVITIES—a statement setting forth the detailed missions assigned to one or more Strategic Services branches, designed to accomplish a given objective, together with a summary of the situation and the general methods of accomplishment of the assigned missions.

d. MISSION—a statement of purpose set forth in a special program for the accomplishment of a given objective.

e. OPERATIONAL PLAN—an amplification or elaboration of a special program, containing the details and means of carrying out the specified activities.

f. TASK—a detailed operation, usually planned in the field, which contributes toward the accomplishment of a mission.

g. TARGET—a place, establishment, group, or individual toward which activities or operations are directed.

SECRET

<u>h</u>. THE FIELD—all areas outside of the United States in which strategic services activities take place.

<u>i</u>. FIELD BASE—an OSS headquarters in the field, designated by the name of the city in which it is established, e.g., OSS Field Base, London.

<u>j</u>. ADVANCED OR SUB-BASE—an additional base established by and responsible to an OSS Field Base, London.

<u>k</u>. OPERATIVE—an individual employed by and responsible to the OSS and assigned under special programs to field activity.

<u>l</u>. AGENT—an individual recruited in the field who is employed and directed by an OSS operative or by a field or sub-base.

<u>m</u>. PARENT CRAFT—the medium by which personnel and supplies are transported from the base to within Maritime Unit operational distance of their objective.

SECTION II—OPERATIONS AND METHODS

3. *MISSIONS*

<u>a</u>. To conduct clandestine ferrying.

<u>b</u>. To conduct maritime sabotage.

<u>c</u>. To provide military tactical assistance.

<u>d</u>. To conduct special training by Maritime Unit.

4. *CLANDESTINE FERRYING*

<u>a</u>. GENERAL—Penetrations into and departures from enemy areas by water will be the specific responsibility of MU. The responsibility essentially will be to effect the transfer of personnel, supplies, and communications from water to land and land to water. Such ferrying which will normally be clandestine may be considered in two stages: approach and departure by parent craft, transfer to and from parent craft.

<u>b</u>. APPROACH TO ENEMY SHORE—This can be by a parent craft of sufficient range and other characteristics necessary to get within small boat or swimming distance of enemy shore. Parent craft may be submarine, de-

stroyer, torpedo boat, or other conveyances. Parent craft may be detailed by U.S. Navy or other Allied armed forces either for specific tasks or regular operations. Parent craft may also be native or other vessels acquired by OSS.

c. TRANSFER OF PERSONNEL AND MATERIEL TO AND FROM PARENT CRAFT TO SHORE—This may be by swimming, surfboard, rubber boat, dinghy, or other small surface craft.

5. *MARITIME SABOTAGE*

Maritime sabotage against enemy shipping and shipping installations in harbors, roadsteads, canals, and rivers, will be executed with limpets and other special underwater demolitions and with standard demolitions. Special Maritime Groups of swimmers are trained to conduct underwater sabotage. However, MU personnel will also participate in maritime sabotage by ferrying demolitions parties to targets or target areas.

6. *MILITARY TACTICAL ASSISTANCE*

a. GENERAL—Where unique techniques and abilities of MU (such as underwater approach and clandestine ferrying and maritime sabotage) are required by a military commander in his theater, such aid by MU shall be furnished as requested of OSS by the theater commander.

b. SPECIAL TACTICAL AIDS—MU sections, when adequately manned at the theater base, may render the following clandestine aid to military operations: (1) hydrographic and beach reconnaissance; (2) establishing navigation aids, especially close to shore; (3) infiltration and exfiltration of personnel.

7. *SPECIAL TRAINING BY MARITIME UNIT*

a. GENERAL—MU will assist Schools and Training Branch by providing instructors and equipment for the training of other OSS personnel and military personnel in special MU techniques, upon request.

b. OSS PERSONNEL—Where operatives or agents have to be infiltrated or exfiltrated by water, they will be

SECRET

trained (usually in the theater) by S&T to enable them to effect the transition from water to shore and vice versa. MU will provide instructors and equipment to assist in such training. Other MU techniques will be taught to OSS personnel of other branches as required for their special tasks.

c. MILITARY PERSONNEL—Where specific MU techniques and equipment are of special use to military commanders in their theater and where training in MU techniques is requested by the military commander through OSS, the MU Section in the theater will provide instructors and equipment to assist S&T in such training.

8. *EQUIPMENT*

a. Specially designed equipment for use under water and on the surface includes self-contained breathing devices, motor propelled surfboards, swim suits, swim fins, two and eight place kayaks, depth gauge, underwater luminous compass, underwater flashlight, electric waterproof motor for use on surf boards and rubber boats. Detailed descriptions of this special equipment are given in a secret pamphlet "Underwater Operations" prepared for the Maritime Unit, December 1943.

b. Standard military and OSS demolitions are used. A principal type is the limpet; the OSS magnetic type and the "pin up" limpet. Military equipment and supplies, such as rations, clothing, small arms, ammunition, and the like will be supplied from U.S. Army or Navy sources in the theaters. Special OSS explosives and equipment will be supplied by Services Branch, OSS.

SECTION III—ORGANIZATION AND PLANNING

9. *BRANCH AND FIELD BASE ORGANIZATION*

a. WASHINGTON—The Chief of MU Branch, Washington, is directly responsible to the Deputy Director, SSO for the carrying out of MU operations. He is assisted by a Deputy Chief, an Operations Officer, a Supply Officer, and a Personnel Officer. Liaison in Washington with

British Commander Combined Operations is maintained through appropriate officers of that nation in contact with the Chief, MU, or any representative designated by him.

b. ORGANIZATION OF FIELD BASES—

(1) The organization of MU at OSS field bases will vary in accordance with local conditions and requirements, but generally they will reflect the structure of the MU Branch, Washington.

(2) The MU Section of an OSS field base is headed by a Chief who is responsible to the Strategic Services Officer.

(3) The Operations Officer of the MU Section of an OSS field base is responsible for planning and coordination of operations with naval vessels detailed to OSS tasks. In the case of naval units, they will be administratively and operationally under the Navy.

(4) All activities of a field base in a theater of operations are under the control and direction of the theater commander.

10. *PLANNING AND ORGANIZATION FOR OPERATIONS*

a. The approved OSS over-all and special programs establish the objectives and missions for MU. Operational plans are developed by MU in the field in conformity with the approved special programs.

b. The MU section in the field assembles personnel and equipment required to accomplish specific missions based upon operational plans developed in accordance with approved special programs.

c. All MU plans and operations are coordinated with the activities of other branches by the Chief of the MU Branch in Washington, and by the Chief of the MU sections at the various field bases. MU advises and assists other branches on any project with maritime phases.

d. MU Branch in Washington is to be kept fully informed of all MU plans and projects for operations originating in the field.

SECRET

SECTION IV—PERSONNEL

11. *REQUIREMENTS*

a. The duties of personnel selected for MU activities divide into four general types:

(1) Staff work at the branch in Washington or in the MU Section at a field base.

(2) Assisting S&T in instruction in special MU techniques.

(3) Maritime operations.

(4) Special underwater swimming activities.

b. For all of these types of activities personnel should be such that the MU special training can be assimilated and employed effectively. For the first three types of duties it is important that the personnel have seafaring experience, particularly with small boats. For the special underwater swimming activities, exceptional swimming ability is a specific requirement.

c. The principal sources for the types of personnel required for MU activities are the U. S. Navy, Marine Corps, and Coast Guard. Competent personnel with special skills are also taken from the Army and civil life.

12. *RECRUITING*

Personnel for MU activities is secured through the regular OSS channels. Requests for personnel are submitted to the OSS Personnel Procurement Branch. This branch makes all arrangements for procuring Army and civilian personnel and forwards requests for Navy, Marine Corps, and Coast Guard personnel to the Naval Command, OSS.

SECTION V—TRAINING

13. *BASIC TRAINING*

a. The basic training for all MU personnel includes the following subjects:

(1) Day and night landings (and reembarkations) through surf.

(2) Swimming in surf and under water.

(3) Handling and maintenance of small boats (rubber boats, kayaks, caiques, etc.)

(4) Navigation, piloting, seamanship.

(5) Reading of maps, charts and aerial photographs.

(6) Hydrographic and beach reconnaissance.

(7) Maritime sabotage instruments and methods.

(8) Harbor and beach defenses.

(9) Demolitions.

(10) Small arms (sub-machine guns, pistol, carbine, rifle, MG).

(11) Operation and simple maintenance of outboard and marine motors.

(12) Operation and care of special MU underwater and surface gear.

(13) Signaling.

(14) Hand-to-hand combat.

(15) Types and designs of ships.

(16) Geography of area of operations.

b. For all types of MU personnel recruited in the U. S., basic training in all subjects is given in the U. S.

c. Personnel recruited overseas are given basic training and specialized training at field schools established in the various theaters.

d. MU instruction for special courses in MU techniques is made available especially in the field to other branches of the OSS and on request to military and naval personnel not assigned to OSS.

14. *SPECIALIZED TRAINING*

a. "Operational Personnel" are specially trained and equipped for special duties such as clandestine ferrying, maritime sabotage, and military tactical assistance. Such advanced training is normally given by MU instructors at field bases.

b. "Special Maritime Groups" of swimmers are given intensive training in underwater swimming (normally a minimum of six months). They are organized and trained specifically for underwater operations and therefore should be used only for tasks for which this special training is required.

SECRET

APPENDIX "A"

EXAMPLES OF TYPICAL MU OPERATIONS

1. *INTRODUCTION*

No attempt is made herein to assess the reasoning and considerations which must precede the assignment of a task to a particular section of an OSS Field Base. This Appendix will serve to illustrate several typical Maritime Unit operations.

2. *MARITIME SABOTAGE (1)*

a̲. PROBLEM—It is desired to attack an enemy vessel moored in the channel of a hostile harbor.

b̲. SOLUTION—Task is assigned to Maritime Unit Section of OSS Field Base, since weighing of all factors concerned indicates that underwater sabotage attack presents greatest likelihood of success.

(1) *Personnel*

Since underwater swimming is required, two Special Maritime Group (SMG) men are assigned the task.

(2) *Method of Attack*

It is planned that one man will affix limpets to the side of the vessel, while the other will secure plastic charges to the fore and aft anchor cables. Use of lungs, swim suits, depth gauges, compasses and fins is required.

(3) *Penetration of Harbor*

(a) This is executed by parent craft (submarine, or surface vessel, depending upon circumstances assigned by Theater Commander) which transports the SMG men to

(1̲) Rendezvous point where friendly native fishermen may pick them up, secrete them, take them into harbor and return them to rendezvous point after they had finished affixing explosives with twelve-hour time charges under cover of darkness, or

8

(2) Rendezvous point and putting them over the side in inflated surfboard. This would be used to take the men within underwater swimming distance of target, then deflated, secured perhaps to a channel marker so that it may be regained, reinflated by special CO_2 bottle and used to rendezvous with parent craft on return, or

(3) Within actual underwater swimming distance of the target.

3. *MARITIME SABOTAGE* (2)

a. PROBLEM—It is desired to destroy an important lock (or dock, or bridge) in an enemy canal. Heavy guard prevents approach from shore.

b. SOLUTION—Task is assigned to MU Section of OSS Field Base, as underwater approach appears to be only reasonably safe method of attempting attack.

(1) *Personnel*

An MU operative (trained as member of Special Maritime Group), thoroughly conversant with the locality, language, customs of the natives and highly trained in demolition, work is selected.

(2) *Method of Attack*

Attack will be made under water and will require the use of lungs, fins, swim suits, gauges, compass and the handling of explosives and fuses under water.

(3) *Penetration*

Operative is parachuted into locality with his equipment.

4. *CLANDESTINE FERRYING* (1)

a. PROBLEM—It is desired to land an OSS Operational Group on a hostile beach so that they may penetrate inland to contact guerrilla forces.

b. SOLUTION—The task of ferrying is assigned to the Maritime Unit of OSS Field Base.

SECRET

(1) *Personnel*

A group of ten OG's is turned over to the Maritime Unit for several days intensive training in landing through surf. Four Maritime Unit men are assigned responsibility for delivery of OG's.

(2) *Approach*

A suitable parent ship is assigned to this particular task. It transports the OSS men to within several hundred yards off shore of landing point on beach under cover of darkness. Two 8-man kayaks are assembled and put over the side. In each are two MU men, five OG's and equipment. The OG's are landed after one MU man has gone over the side and swum in to assure that reception committee of guerrillas with whom rendezvous has been established are on hand and that landing has been made at correct point. Similar technique is followed to evacuate personnel from beaches.

5. *CLANDESTINE FERRYING* (2)

a. PROBLEM—It is desired to land two native SI agents in an enemy port.

b. SOLUTION—Task of ferrying is assigned to Maritime Unit Section of OSS Field Base.

(1) *Personnel*

One MU operative, operating under cover as a native fisherman is assigned responsibility for task.

(2) *Method of Penetration*

The MU operative is in command of a felucca with a reliable crew of natives. This vessel regularly engages in off shore fishing and delivers catch into nearest port, village or harbor every several days. This craft delivers the two SI agents directly into port of their objective, all personnel concerned being under cover as natives.

6. *CLANDESTINE FERRYING* (3)

a. PROBLEM—It is desired to establish communications with partisan groups on a coastal island which must be approached through enemy-controlled waters. The purpose is to supply them continuously with arms, ammunition, food and medicines.

b. SOLUTION—Task is assigned to Maritime Unit Section of OSS Field Base.

(1) *Personnel*

The MU section has trained a number of natives to act as crews for native caiques with under cover MU operatives as commanding officers. Three such vessels with crews are assigned responsibility.

(2) *Method*

Under cover as fishing boats and native ferries, these vessels accomplish their assigned mission by continuous ferrying of supplies to objective.

SECRET

MU
maritime unit

This booklet has been prepared for use within OSS, particularly by MU Branch for the general orientation of its own personnel. Its primary purpose is to define the basic organization of the Branch, its functions, doctrine, and purpose. It is neither a technical training manual nor a historical record of actual accomplishments.

OFFICE OF STRATEGIC SERVICES

OFFICE OF STRATEGIC SERVICES

is an agency of the Joint Chiefs of Staff, charged with collecting and analyzing strategic information and secret intelligence required for military operations, and with planning and executing programs of physical sabotage and morale subversion against the enemy in support of military operations.

MARITIME UNIT

plans and carries out the amphibious phases of these activities, and assists in the development of the special equipment required. It penetrates enemy coastal areas, introducing operatives and their equipment for maritime sabotage and other OSS operations. Wherever targets for sabotage may be reached by water, access to enemy land areas may be obtained by water, wherever information is required on water approaches and character of shoreline and coast, MU's special techniques in clandestine ferrying, maritime sabotage, and beach and hydrographic reconnaissance are utilized.

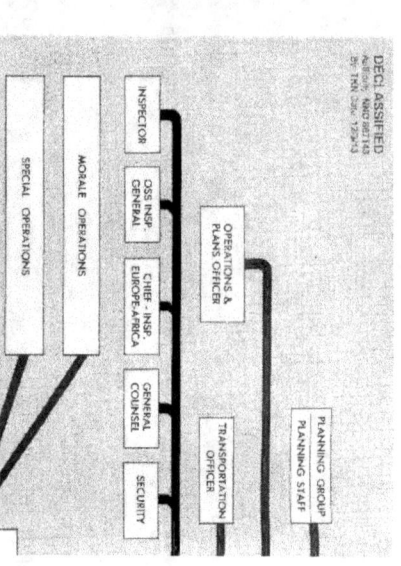

MARITIME UNIT

MU

OFFICE OF STRATEGIC SERVICES

CLANDESTINE FERRYING

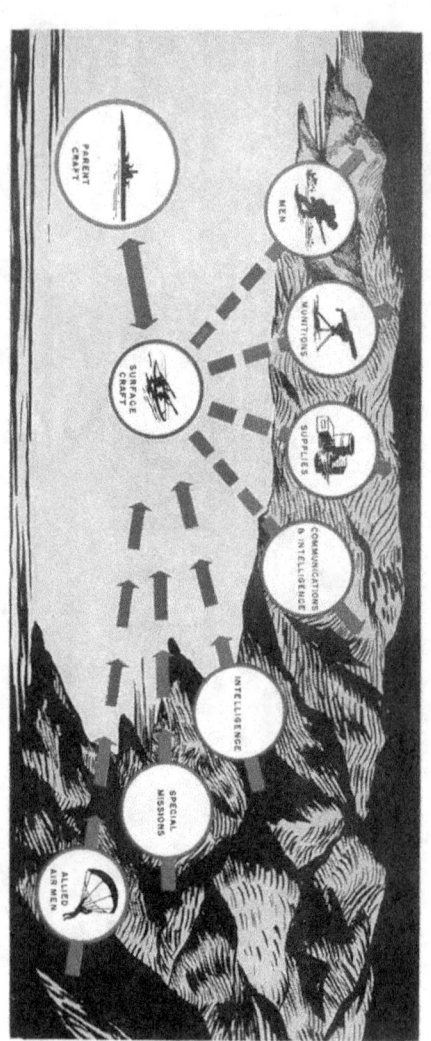

MU OPERATIONS ARE THE AMPHIBIOUS PHASES OF OSS INTELLIGENCE AND SABOTAGE

Men, munitions, supplies, and communications are secretly infiltrated into enemy areas over water, and communications and returning personnel brought out. OSS intelligence or demolition operatives, liaison officers to guerrilla or resistance groups, or special missions from the Theater Commander may be transported. Airmen shot down over enemy-held territory are brought back. Specially equipped operatives may be landed to carry out beach reconnaissance on the character and gradient of beaches and the depths and shallows of the off-shore coast, data of value in planning amphibious assaults.

The parent craft, which may be a submarine, destroyer, or motor torpedo boat, penetrates to within landing distance of the enemy coast. The operatives transfer to small surface craft, surfboard, rubber boat, or kayak, for the trip to and from the shore.

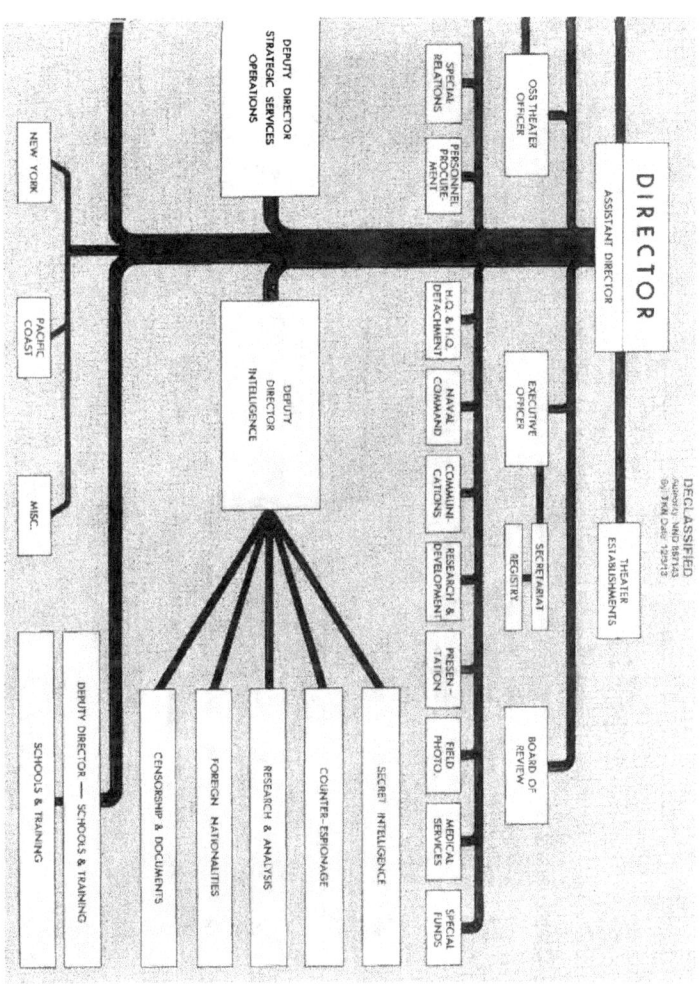

MU PLANNING AND ADMINISTRATION IN WASHINGTON

Strategic services plans and over-all programs are initiated and developed by the Planning Staff in conjunction with the Plans Officer of each branch. The Planning Group integrates the plans with military and naval operations. The Operations and Plans Officer informs the Director of the plans and operations of the branches in the process of development or execution. When the Director has approved a plan, it is forwarded to the Joint Chiefs of Staff for consideration and submitted to the Theater Commander for final approval.

MU Headquarters is chiefly concerned with recruiting and training personnel and with procuring the special equipment needed for operations in all theaters.

For the Central Pacific Theater, MU Headquarters not only fulfills these functions but also originates specific operational plans.

MARITIME SABOTAGE

Swimmers, especially trained in the use of underwater equipment and techniques, attack enemy shipping and port installations. Ferried close to their objectives in a small boat or raft, they swim under water, carrying an explosive charge.

Either Limpets or standard waterproofed demolition charges are used for attacks on the hulls of enemy vessels. Fixing the charge to the target, the operative returns unobserved under water to the ferrying craft. The charge is detonated by a time delay.

Expert underwater swimmers also perform offshore hydrographic reconnaissance.

MU TRAINING

To train men for its specialized operations, MU conducts its own training program under the general supervision of Schools & Training Branch. Men selected as operatives must be highly qualified in swimming and small boat handling. After intensive instruction at basic and advanced schools the men are sent to field bases overseas where training is continuously carried on between missions to keep them in first-class physical condition and familiar with the latest developments in equipment and technique.

OPERATIONAL PERSONNEL infiltrate and exfiltrate Special Maritime Groups, and personnel and supplies for other OSS branches, to and from coastal areas of enemy-occupied territory. They also instruct agents and native operators in the use and handling of small boats and in beach reconnaissance.

SPECIAL MARITIME GROUPS carry out maritime sabotage and offshore hydrographic reconnaissance.

RECRUITING
- NAVY
- MARINES
- COAST GUARD
- ARMY
- REGULAR OSS CHANNELS

→ CANDIDATE APPRAISAL BOARD →

- OPERATIONAL PERSONNEL
- SPECIAL MARITIME GROUPS

MU FIELD BASE ORGANIZATION IN SEAC

varies with the operational opportunities and requirements of the theater.

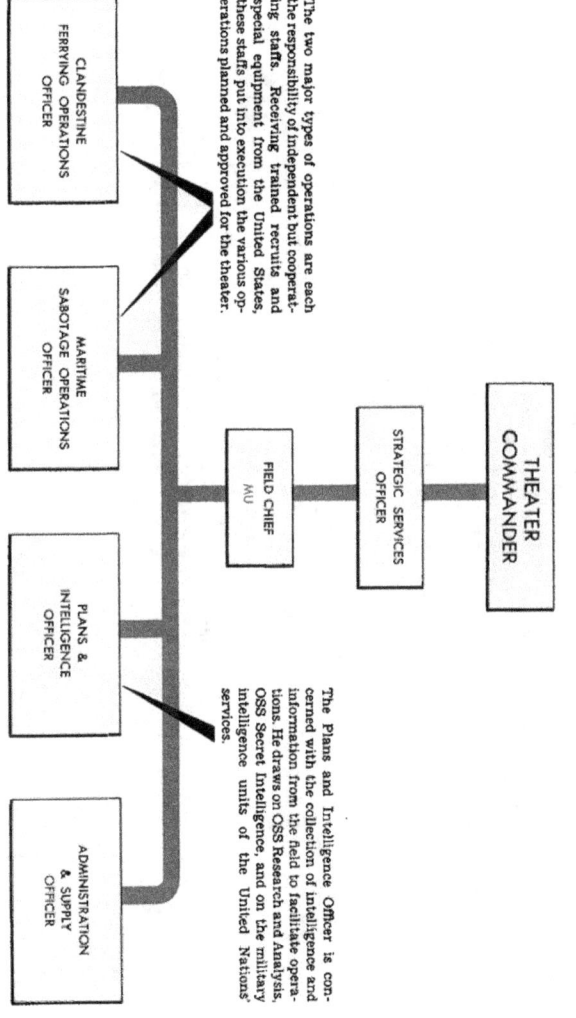

The two major types of operations are each the responsibility of independent but cooperating staffs. Receiving trained recruits and special equipment from the United States, these staffs put into execution the various operations planned and approved for the theater.

The Plans and Intelligence Officer is concerned with the collection of intelligence and information from the field to facilitate operations. He draws on OSS Research and Analysis, OSS Secret Intelligence, and on the military intelligence units of the United Nations' services.

M U   E Q U I P M E N T

Special equipment is required by MU to carry out its various missions. It has actively participated in the development of these specialized

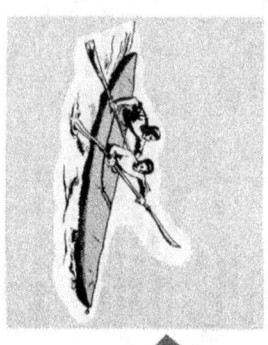

THE TWO-MAN KAYAK is a collapsible, portable boat made of a plywood frame covered by rubberized canvas. Assembled, the boat weighs 104 lbs, is 16½ ft. long, has a beam of 34 in. and a depth of 13 in. It can be assembled by two men in less than five minutes.

THE EIGHT-MAN KAYAK is built on the same principles as the two-man kayak. It weighs 245 lbs, is 24 ft. long and 19 in. deep. This boat has detachable outriggers for the addition of two outboard motors.

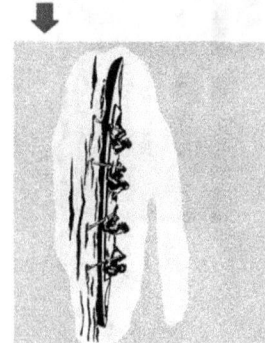

The shaped plywood frames for both kayaks are fitted together by lengths of metal pipe which also serve to stabilize the craft. The entire framework fits into a one-man haversack.

The rubberized canvas hull is made of tough fabric which can be repaired with rubber patches and cement and will not run when punctured.

The double-bladed paddles are collapsible to facilitate packing and transportation.

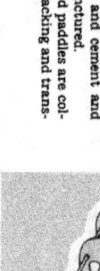

6 WEEKS BASIC INSTRUCTION

OPERATIONAL PERSONNEL
Maintenance and handling of small boats and motors.
Operation and care of special MU equipment.
Navigation, piloting, seamanship. (Hydrographic and beach reconnaissance.)
Reading of maps, charts, and aerial photographs.
Basic OSS operational instructions.

SPECIAL MARITIME GROUPS
Preliminary swimming instruction and diving.
Maintenance and handling of small boats and motors.
Operation and care of special MU equipment.
Demolitions.
Maritime sabotage instruments and methods.
Basic OSS operational instructions

→ **CLANDESTINE FERRYING**

8 WEEKS ADVANCED INSTRUCTION

Advanced underwater swimming.
Hydrographic and beach reconnaissance.
Operational problems.
Communications.
Harbor and beach defenses.
Navigation.
Geography of operational areas.

→ **MARITIME SABOTAGE & BEACH RECONNAISSANCE**

SURFACE CRAFT

devices with the Research and Development Branch of OSS. The small surface craft are designed for clandestine approach to enemy shores.

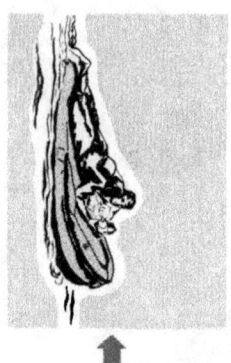

THE TWO-MAN SURFBOARD is a pneumatic rubber surfboard, 10½ ft. long, 3 ft. 7 in. wide, weighing about 310 lbs. A compressed air cylinder inflates it in a few minutes. A battery-driven ¾ horsepower motor with a speed of 5 knots and a range of 10 miles can be attached. The surfboard carries two men and their equipment, the equivalent of 900 lbs.

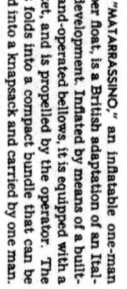

THE "MATARASSINO," an inflatable one-man rubber float, is a British adaptation of an Italian development. Inflated by means of a built-in hand-operated bellows, it is equipped with a pocket, and is propelled by the operator. The float folds into a compact bundle that can be fitted into a knapsack and carried by one man.

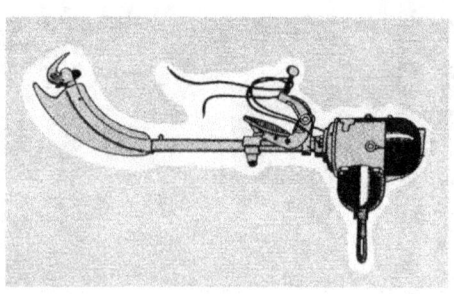

THE ELECTRIC MOTOR is driven by a 12-volt, 120-ampere battery. Silent in operation, it is used as motive power for the surfboard.

MU UNDERWATER EQUIPMENT

Underwater approaches to enemy shipping and installations for sabotage have been made feasible through the development and use of specialized equipment.

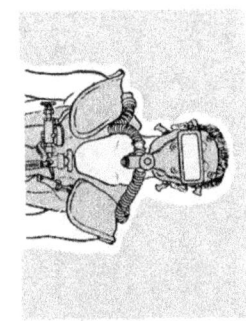

THE LAMBERTSON UNIT is a completely self-contained breathing device which enables a swimmer to approach his target beneath the water surface without leaving a wake, breathing bubbles, or other traces of his presence or movements. The face piece provides excellent visibility, and all controls and gauges are accessible without hampering the swimmer's motions. This unit allows a swimmer to spend an hour under water, and travel more than a mile.

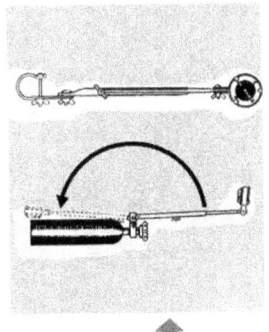

THE UNDERWATER COMPASS is a standard waterproof compass fitted into a holder which is attached to the Lambertson Unit. A joint enables the swimmer to raise the compass to eye level or push it down flush to his body. It does not interfere with swimming in either position.

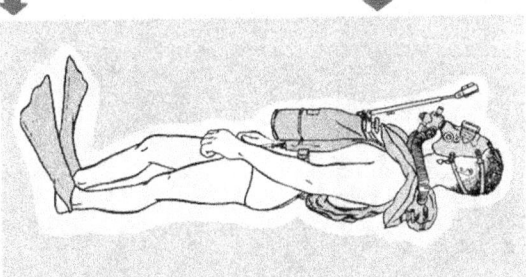

SWIM FINS. The large Swim Fin is the improved OSS design. Its base is the standard swim fin on which an extra length of live rubber has been vulcanized. The proper use of swim fins doubles a swimmer's speed and range.

25

MU OPERATIONS

Representative of MU operations are those carried out in the Mediterranean theater.

Extensive and highly successful clandestine ferrying operations have been carried out in the Aegean Sea. OSS personnel and supplies have been transported to Greece in support of native resistance groups and guerilla forces. Fleets of caiques ply between Cyprus and secret bases in Turkey. From these bases smaller craft operate, making pinpoint landings at night on the Greek coast. Refugees and downed fliers, rescued and brought to evacuation points by Greek guerrillas, have been picked up and brought to Cyprus.

MU plans and supervises the operations of an amphibious group of the Italian Navy in the Tyrrhenian Sea and the Adriatic Sea. Specializing in demolition attacks and shore sabotage, the unit has a successful record under MU control.

Clandestine ferrying operations across the Adriatic Sea in support of Yugoslav guerrillas have been conducted by MU from Bari. A ship repair service has also been conducted there for OSS Special Operations Branch.

A few infiltrations along the coast of Southern France have been carried out by MU units based on Corsica.

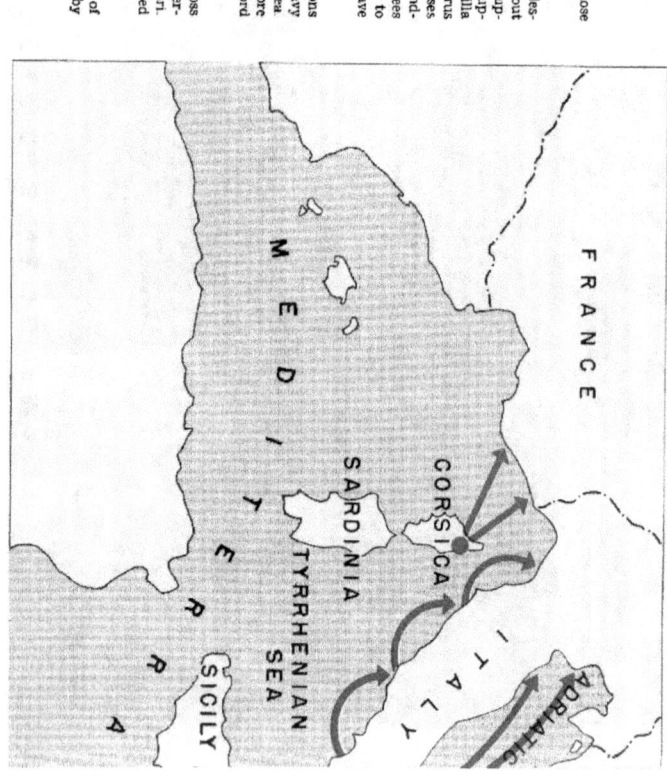

SPECIAL EQUIPMENT

Special light-weight waterproof devices enable the beach reconnaissance operative with a minimum of apparatus to make and record accurate observations of beach composition, gradient, and depth.

THE LIMPET is an explosive weapon designed for use against enemy merchant vessels. The explosive charge is contained in a waterproof plastic case equipped with strong magnets for adhesion to the metal plates of the target.

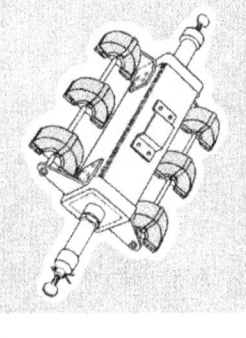

The Limpet is most effective when used about 5 ft. below the water surface, its charge being sufficient to blow a hole about 25 sq. ft. in a merchant vessel. The Limpet may be attached under water either by a swimmer or from a small boat with the aid of an extendable placing rod. A range of time delays permits the operator to make his escape.

THE PIN-UP GIRL is essentially the same explosive weapon as the Limpet, but the method of attaching the container to the target has been changed. A cartridge-driven pinning device which will penetrate wood or steel hulls has been substituted for the magnets of the Limpet case.

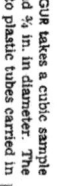

THE BEACH SAND AUGUR takes a cubic sample of sand 8 in. long and ¾ in. in diameter. The sand is transferred to plastic tubes carried in a bandolier.

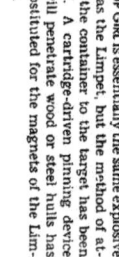

UNDERWATER PAD AND PENCIL. The writing pad, mounted on a strong leather wrist strap, is made of white "ivorine" cards deeply grooved in one inch squares to facilitate the tabulation of soundings in the dark. The underwater pencil is a modification of the vest-pocket flashlight. A chinagraph pencil has been added. The lens in the flashlight is movable, allowing the operator to regulate the amount of light transmitted.

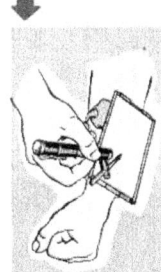
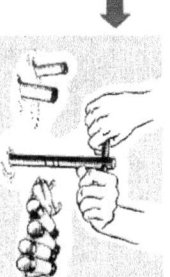

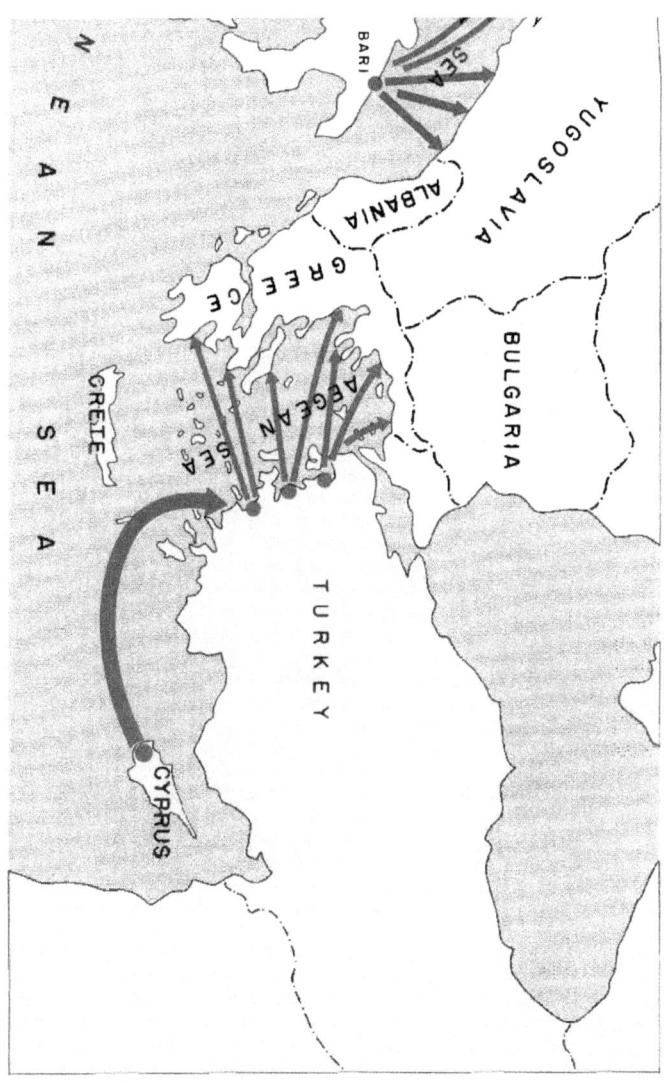

LITHOGRAPHED IN THE REPRODUCTION BRANCH, OSS

www.ingramcontent.com/pod-product-compliance
Lightning Source LLC
Chambersburg PA
CBHW071221240526
45470CB00018B/2192